Children's Science Library

COMMUNICATION

Author & Illustrator

A.H. Hashmi

Editor

Rajiv Garg

Published by:

F-2/16, Ansari road, Daryaganj, New Delhi-110002
☎ 23240026, 23240027 • *Fax:* 011-23240028
Email: info@vspublishers.com • *Website:* www.vspublishers.com

Branch : Hyderabad
5-1-707/1, Brij Bhawan (Beside Central Bank of India Lane)
Bank Street, Koti Hyderabad - 500 095
☎ 040-24737290
E-mail: vspublishershyd@gmail.com

Distributors :

- **Pustak Mahal**®, Delhi
  J-3/16, Daryaganj, New Delhi-110002
  ☎ 23276539, 23272783, 23272784 • *Fax:* 011-23260518
  E-mail: sales@pustakmahal.com • *Website:* www.pustakmahal.com
  Bengaluru: ☎ 080-22234025 • *Telefax:* 080-22240209
  Patna: ☎ 0612-3294193 • *Telefax:* 0612-2302719
- **PM Publications**
  - 10-B, Netaji Subhash Marg, Daryaganj, New Delhi-110002
    ☎ 23268292, 23268293, 23279900 • *Fax:* 011-23280567
    E-mail: pmpublications@gmail.com
  - 6686, Khari Baoli, Delhi-110006
    ☎ 23944314, 23911979
- **Unicorn Books**
  Mumbai :
  23-25, Zaoba Wadi (Opp. VIP Showroom), Thakurdwar, Mumbai-400002
  ☎ 022-22010941 • *Telefax:* 022-22053387

ISBN 978-93-814483-4-2
Edition 2011

Printed at : Param Offset Okhla, Delhi

CONTENTS

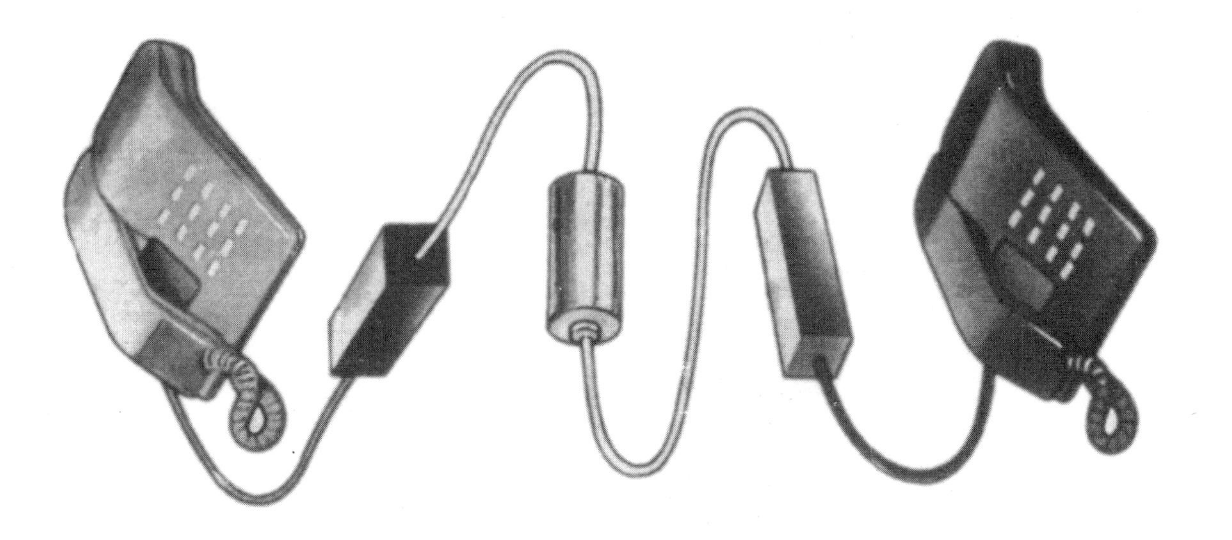

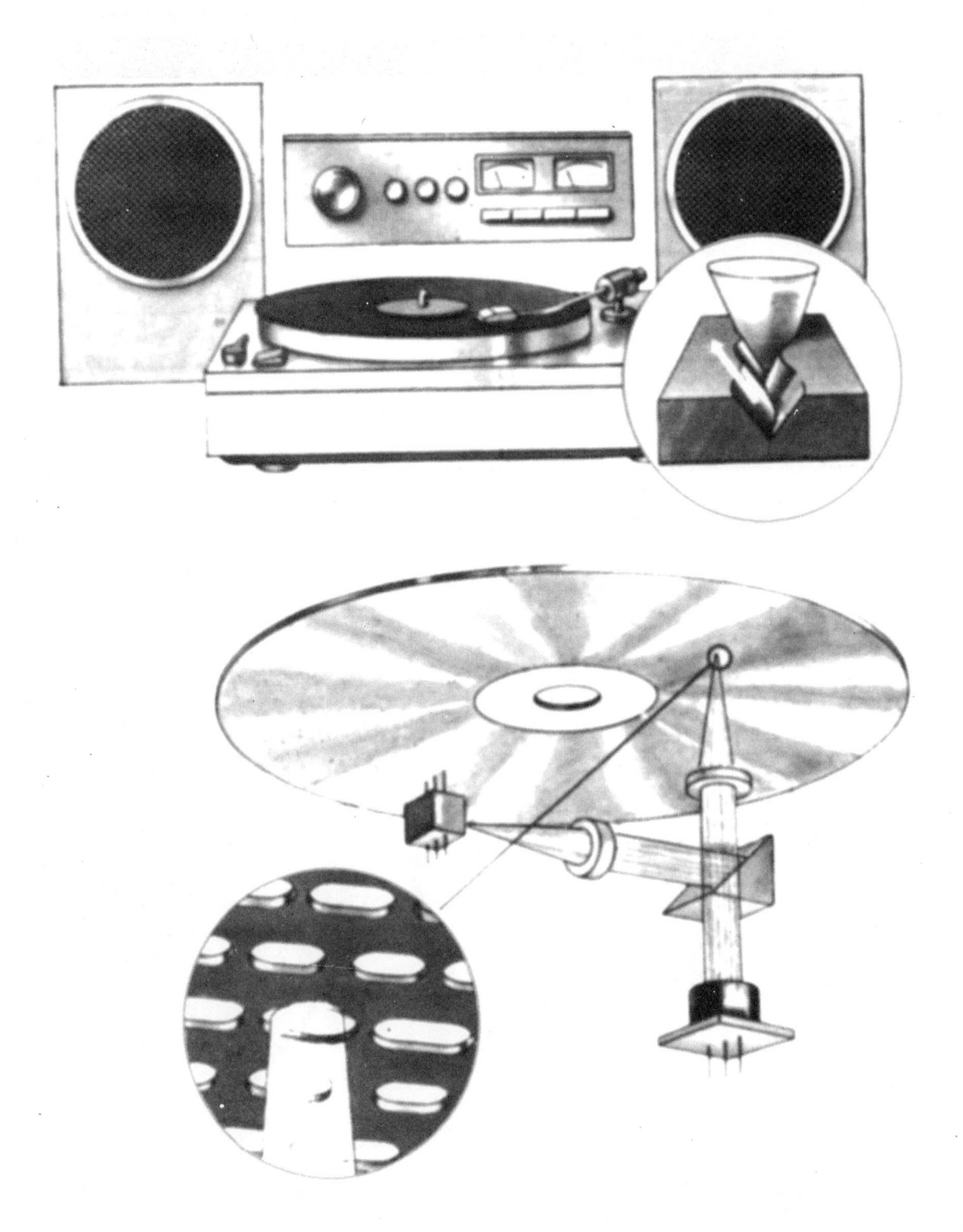

COMMUNICATION

The word 'communication' means exchange of information. Animals communicate with each other by special sounds. In olden times, man could communicate by shouting or blowing a horn or beating a drum or flashing a light. Man gradually developed the art of talking and writing and started expressing his complex thoughts and information through language and writing. In the Mughal period, pigeons were used as carriers of messages. In the early 16th century, the postal system had begun. But the riders were used till 1830 for carrying letters and parcels to distant places.

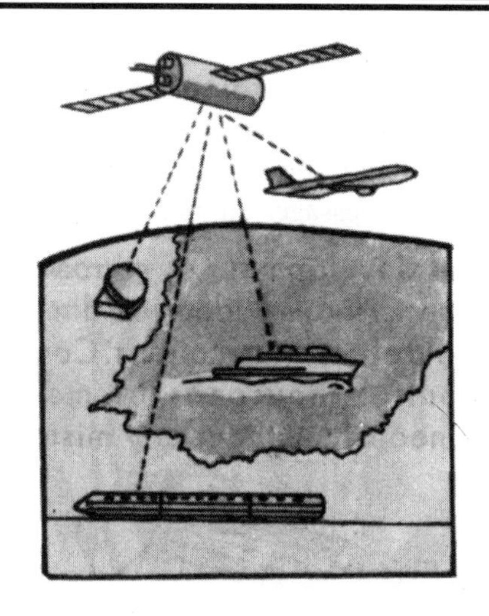

For distant communication, nowadays communication satellites are used

Communication took a new turn with the development of technology. In 1837, Cooke and Charles of England and Samuel Morse of America developed the *electric telegraph*. The information was sent in the form of electrical signals or codes through a cable by this instrument. The dots (short clicks) and dashes (long clicks) were used to represent the letters of the alphabets. In 1876, Alexander Graham Bell invented the *telephone* by which it became possible to communicate by voice over long distances through cables. In 1894, Guglielmo Marconi of Italy invented the *wireless telegraph* by which messages could be sent across a long distance without wire. The efforts of these great scientists made it possible to send or receive messages from any part of the globe within no time.

The electric telegraph led to the development of the *telex system* while the wireless telegraph gave birth to the *radio*. Telephone networks have advanced very much. Besides messages, we can also send pictures and documents by fax.

Radio waves have played an important role in the communication systems. Radio waves are

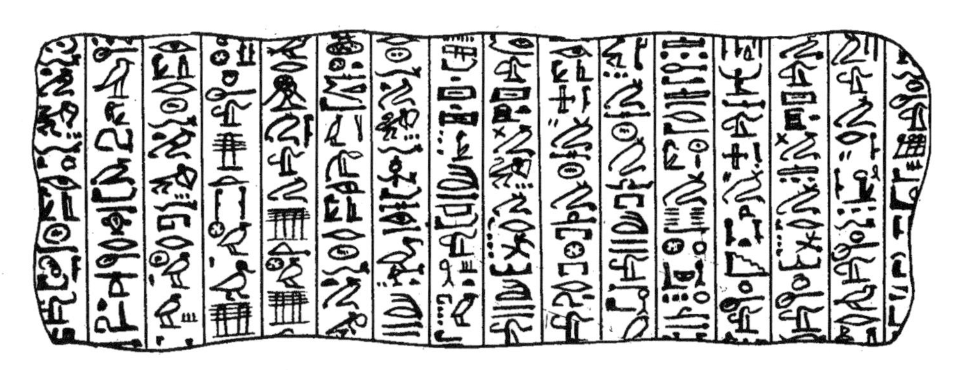

Around 5,000 years back, messages were conveyed by means of inscribing pictures on rocks

used in radio, telex, telephone, and television systems. Very short radio waves or *microwaves* are used to send messages across sea using *satellites*.

The development of *microelectronics* and *computers* have added new dimensions in the field of communications. Computers can control thousands of messages simultaneously without any mistake.

The fax machine can send and receive both words and pictures

In future, everyone will have a compact telephone

Books, newspapers, radio, television, cinema, etc. are other well-known means of communications. Through these means, we can receive information from any part of the world within no time.

PRINTING

Printing is a process in which a large number of copies of text and pictures is produced on paper in a short time. This art was developed in the sixth century in China. After about 500 years, movable type was also invented in China. From China, this technique became popular in European countries. In the fifteenth century, Johannes Gutenberg of Germany developed the printing process in Europe. In 1476, William Caxton introduced the first printing press in London.

Platen press, which used to print sheet by sheet, was replaced by *rotary press*. A rotary press continuously prints a paper roll. The printing which uses movable type is called *letter press*. In *lithography*, smooth plate is used. Text or picture is taken on a greasy surface, while the remaining part of plate is covered with grease-repelling material. When greasy ink is applied onto the plate, it adheres on the greasy parts only.

Offset printing is a modified form of lithography. The main parts of an offset machine are: a plate cylinder, a blanket cylinder and an impression cylinder. The plate cylinder marks an impression on rubber and it prints on paper.

The invention of computer has brought many improvements in printing. Nowadays, edition of text, printing and designing of page is done with the help of computers. The publishers of this book, Pustak Mahal have made all the pages of this book by using *DTP (Desk Top Publishing)* based computers.

Word Processing

The type and designing of any text, newspaper or book can be done very fast and easily in required size in a *word processor* with the help of a computer. A word processor makes use of electronic system by which written matter can be edited on its own. We can alter the sequence

The German, Johannes Gutenberg established the typing process in Europe

A page from Gutenberg's Bible

of words and sentences, even delete or add words without typing the total matter again.

Whatever you type on the keyboard gets displayed on the *visual display unit VDU* like a TV screen. Operators can do any change or correction in the text by using the *editing key* and *movable pointer*. Text gets stored in the memory of the word processor, which can be arranged accordingly by looking at the screen. Operators can store the text permanently on disk or tape, which can be used afterwards whenever needed. Word processor has a link with the electronic printer, by which a typed copy of text stored in memory comes out.

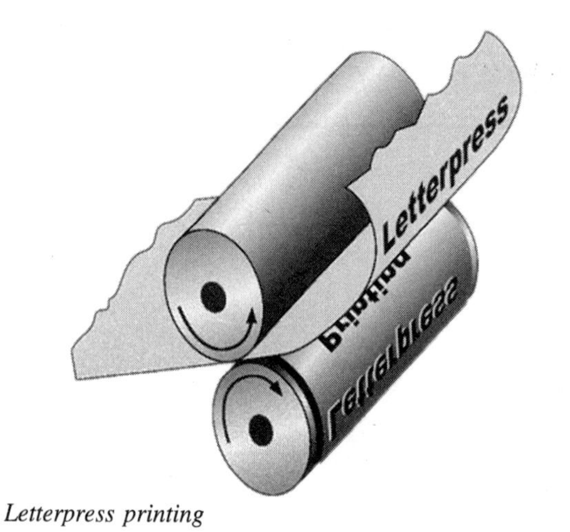

Letterpress printing

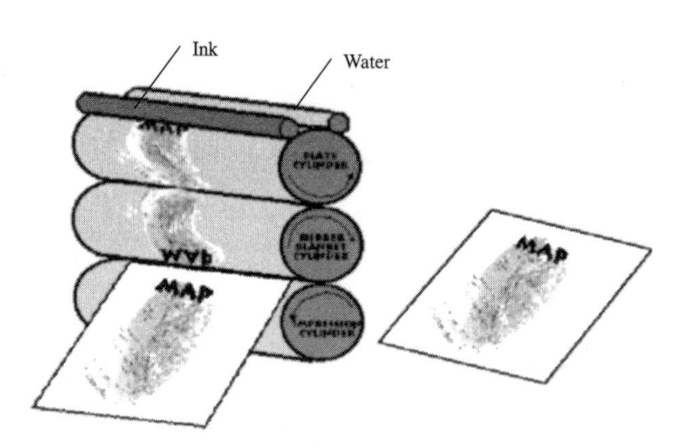

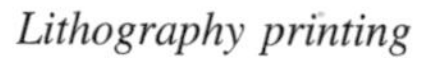

Lithography printing

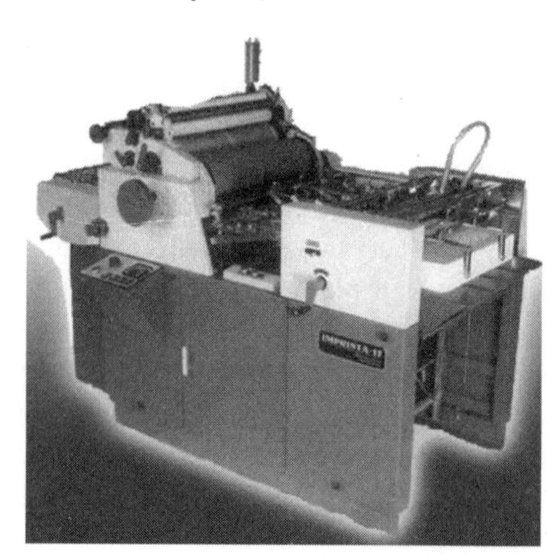

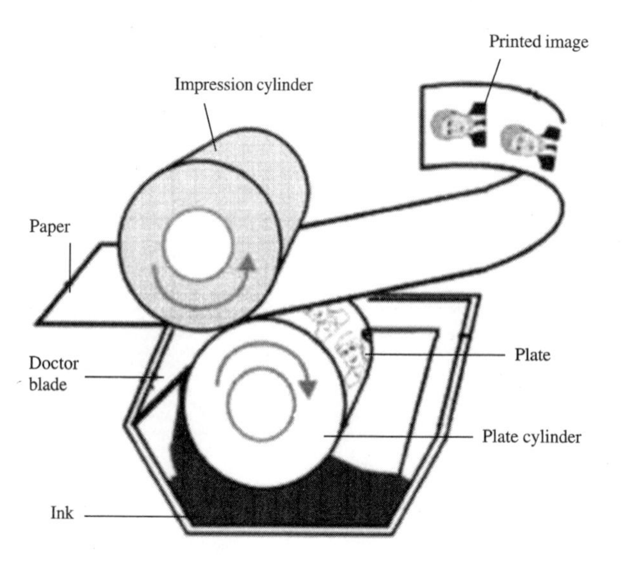

Gravure printing

Offset machines

Word processor

Word processor also helps to set the text into pages. Line length, spacing, indentation, margins, etc. can be set according to will. When it is printed, the machine can arrange the text on its own, if a word is longer and is coming out of the line, the word processor brings this word in the next line and adjusts the spacing of the first line. In the same way, the size of the letters etc. can also be changed.

PHOTOGRAPHY

The technique of recording the image of an object on a *photosensitive* film with the help of a camera is called *photography.* In the beginning, metal or glass plates were used to records the images. *Plastic roll films* were developed in 1889. In the middle of the nineteenth century, several other chemical methods were invented for developing images in addition to Daguerreutype and calotype negative-positive methods. The negative-positive method is in use even today.

L.J.M. Daguerre (1787-1851)

All the cameras have a lens by which an inverted image of the scene is formed on the *photofilm.* A photosensitive layer of silver bromide is coated on photofilm. This is called *exposing.* After exposing, the film is developed in chemicals and then fixed. In this way, a permanent negative of a scene is made on the film. From this negative, prints of required size can be made on bromide paper using an enlarger.

In coloured photography, two types of films are used. The first one is colour-reversal film, on which colour positive, slide or transparencies are made. The other one is colour-negative film, which is used for making colour prints. Both of these films are sensitive to the three primary colours i.e. blue, green and red. In one colour film, we have three layers of photosensitive emulsion. The first layer is sensitive to blue colour, second for green and third for red. Each emulsion layer absorbs or subtracts a definite amount of light. This is called *subtractive process.* The coloured films are developed in a special developer and prints of desired size are made from them.

RADIO

Radio is one of the most effective means of sending messages, news, music, etc. to distant places. In this process, radio waves are used, which are a type of *electromagnetic waves.*

The credit of inventing radio goes to Guglielmo Marconi of Italy. In 1901, Marconi succeeded in sending radio from England to Newfound land.

The frequency of the radio waves used in radio broadcast lies between *150 kilohertz* and *30,000 megahertz*. The messages are transmitted through a *transmitter* which is connected to the *radio station*. Transmitter consists of instruments like *microphone*, *modulator*, *amplifier*, *oscillator*, *antenna*, etc. The person in the radio station speaks in front of a *microphone*, which converts sound waves into electrical signals. These signals are mixed with carrier waves in a modulator, amplified and transmitted.

The modulation of radio waves is of two types: *amplitude modulation* and *frequency modulation*. The messages transmitted by transmitter are received by the aerial of our radio set. A radio set converts the electrical signals into sound and we hear the original programme.

Guglielmo Marconi (1874-1937)

Radio is used in telecommunications, navigation, etc. The radio services started in 1936 in our country. Today, there are more than 100 radio stations in our country.

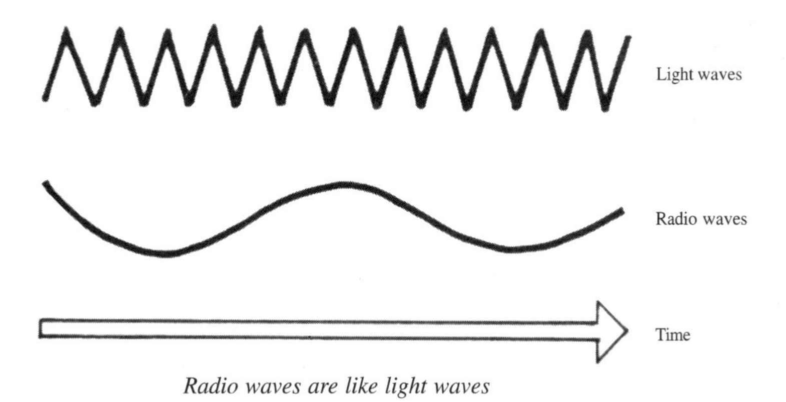

Radio waves are like light waves

RADAR

The instrument used to detect the position and distance of an object is called *radar*. Bad weather, darkness and smoke do not effect the performance of a radar. Radars are mostly used to determine the position, speed and distance of aircrafts and ships.

An echo is produced when sound waves are reflected after striking an object. Similarly, radio waves produce an echo when they get reflected after striking a surface. The invention of radar is based upon the same principle. This is called the *echo principle*.

The word RADAR stands for Radio Detection and Ranging. The first successful radar was developed by Robert Watson Watt and his colleagues in 1930 in Britain.

The *transmitter* of radar sends short pulses of high frequency from a rotating antenna. These pulses are reflected back after striking an object and are received by a *receiver*. The *screen* of the radar shows the position of the object. The time taken by the wave in going from the transmitter of radar to the object and then after reflection back to the receiver is determined. When this interval of time is multiplied by the velocity of light, we get double the distance of the object from the radar. The display unit fixed in the radar shows half of this distance. Besides aeroplane and ship control, radar has played an important role in missile control, transport control, army, weather forecast, spacecraft control, etc.

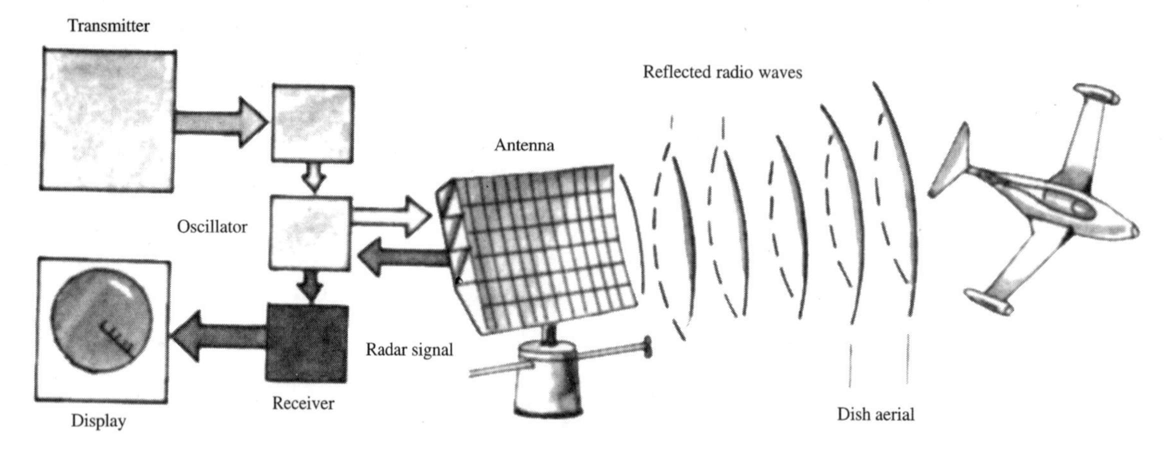

A radar can detect enemy planes

CINEMATOGRAPHY

The moving pictures of cinema are based on the *persistence of human vision*. In fact, pictures do not move but still pictures appear on the screen one after the other. If we are looking at an object, then even after removing the object, its image persists in our eyes for about 1/30 of a second and it gives an impression that the object is there before our eyes. When pictures are shown very fast, one after the other on the screen, our mind is under the misapprehension that the pictures are moving. The second picture comes into the brain before the mental picture of the first is vanished. Then after a fraction of a second, a new picture appears which is slightly different from the previous one. In this way, the process continues and we get an *illusion* of moving pictures. The cinema pictures are shown at a speed of 24 scenes per second.

A few scenes from films made by the Lumiere brothers

In a cinema film, we have a long strip of thousands of still pictures, which is wound around a *spool*. The spool is mounted on a *projector*. The projector shows every picture or frame on a screen. Every second, 24 frames or pictures are projected on the screen so that the scene looks alive. Images on the film are recorded by a *movie camera*. This camera works like a still camera, the only difference is that it takes 24 pictures per second on the film. On one side of the film, sound signals are recorded. In the positive print of the film, sound and picture are both synchronized. This is called *synchronization*.

The first cinema or movie was made by Thomas Alva Edison. This instrument was called *kinetoscope*, in which moving pictures were seen through a small hole.

The credit of showing cinema film goes to the Lumiere brothers of France. They made the first cinema projector by which film could be shown to a large number of audience at a time.

In March 1895, they showed their first short film with the help of a camera and projector in their factory. The world's first cinema hall was built in France. Posters were pasted on the walls of the city on which photographs of the Lumiere brothers were printed and 'Lumiere's Cinema' was written in bold letters.

Cinema was started in 1895 and for 34 years silent films were made. The first talkie was made by the Warner brothers of America in 1927. The name of the film was *The Jazz Singer*. In the year 1930, coloured films appeared. After that, cinema developed very fast. The wide screen cinemascope films were made in the middle of this century. Today, we can see real scenes due to the techniques like *cinerama* and *technirama*.

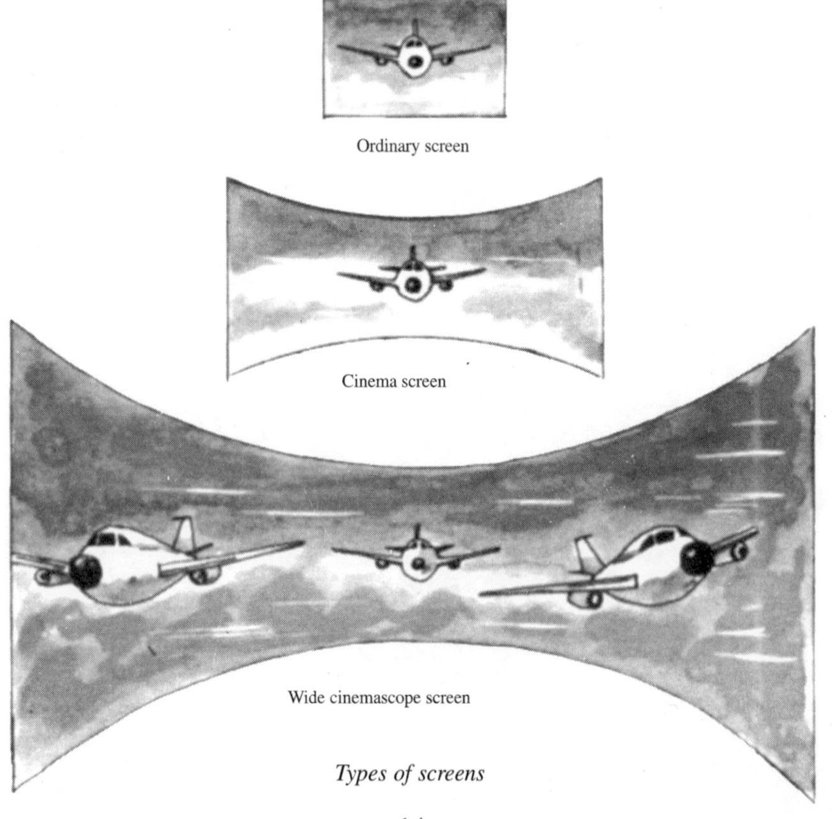

Types of screens

REPRODUCTION OF SOUND

Sound is a kind of energy which travels in the form of waves. It can be recorded on a *tape* or *disk* by converting it into electrical signals. It can also be reproduced and heard from the tape or disk. This whole process is called *sound recording* and *reproduction*. Sound waves are recorded on tapes in the form of *magnetic patterns*. On ordinary record disks, waves are recorded in the form of *grooves*. In a *compact disk*, sound is recorded as spiral patterns of microscopic pit. All these systems of recording sound are different from each other but all require a *microphone* for recording and a *loudspeaker* for reproducing the sound.

Thomas Alva Edison was the foremost person to record sound using a phonograph. The first sentence recorded was –"Mary had a little lamb."

The microphone converts the sound waves into varying electric current. A simple microphone consists of a thin disk or diaphragm connected to a piezo-electric crystal. When this crystal is pressed by sound waves, a feeble electric current is generated. Sound waves cause vibrations in the diaphragm of the microphone. The vibrating diaphragm exerts a little pressure on crystal and, as a result, a feeble electric current is generated. The current produced varies as the amplitude of the vibrations and produces

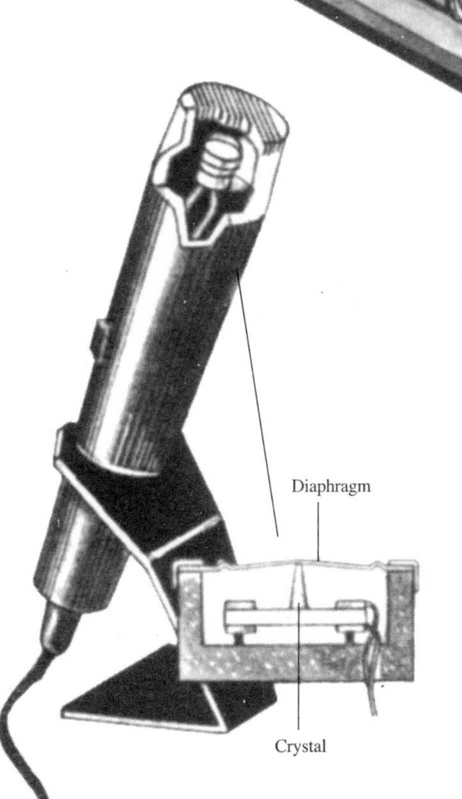

Crystal microphone

Moving-coil loudspeaker

a sound wave pattern which is recorded on tape or disk.

To listen to this sound again, electric current is produced from the disk or tape and this electric current is sent to the loudspeaker. The loudspeaker converts these electrical signals into sound waves. An ordinary loudspeaker has a big cone of paper connected to a coil of wire. The cone is fitted in between the pole pieces of a permanent magnet. When electric current is passed through this coil, a magnetic field is produced which causes vibrations in this cone and sound is produced. In this way, the recorded sound is reproduced and heard.

Record Disc

The credit of recording sound on an ordinary disc goes to the American scientist, Thomas Alva Edison. He invented the *phonograph* in 1877.

In 1887, the production of modern type of disc gramophone started. Commercial disc recording started in 1895.

Record player

If you see the disc with a magnifying glass, you will observe wavy grooves on it. These wavy grooves are formed according to the intensity of vibrations of sound. Sound waves are recorded as an image in these grooves. After recording, the disc is rotated with a needle placed on it. Due to the up-and-down movement of the pin vibrations are produced in it and thus, the original sound gets produced. Today, many developments have been made in the field of recording and reproduction of sound. Nowadays, amplifiers are also used. Mechanical reproduction has been replaced by electrical pickup system. *Long play* and *stereo records* have also developed.

Tape Recorder

Tape recorders were started by Valdemar Poulsen in 1899. Sound is recorded on a plastic tape which remains wound like a reel in a cassette. Most of the cassette tapes are coated with iron oxide, which is a magnetic material.

In a tape recorder, first of all, electrical signals coming from a microphone are converted into magnetic signals. This is done by a small electromagnet which is called the *recording head*. Tape is passed through the head with the help of a motor. The iron oxide coated on it changes to a magnet by the electrical current generated due to sound waves. In this way, sound is recorded on the tape in the form of magnetic field.

To listen to the sound recorded on the tape, it is again passed through the head. The changing electric current due to the head resembles the electric current of the microphone at the time of recording. This electric current is amplified by

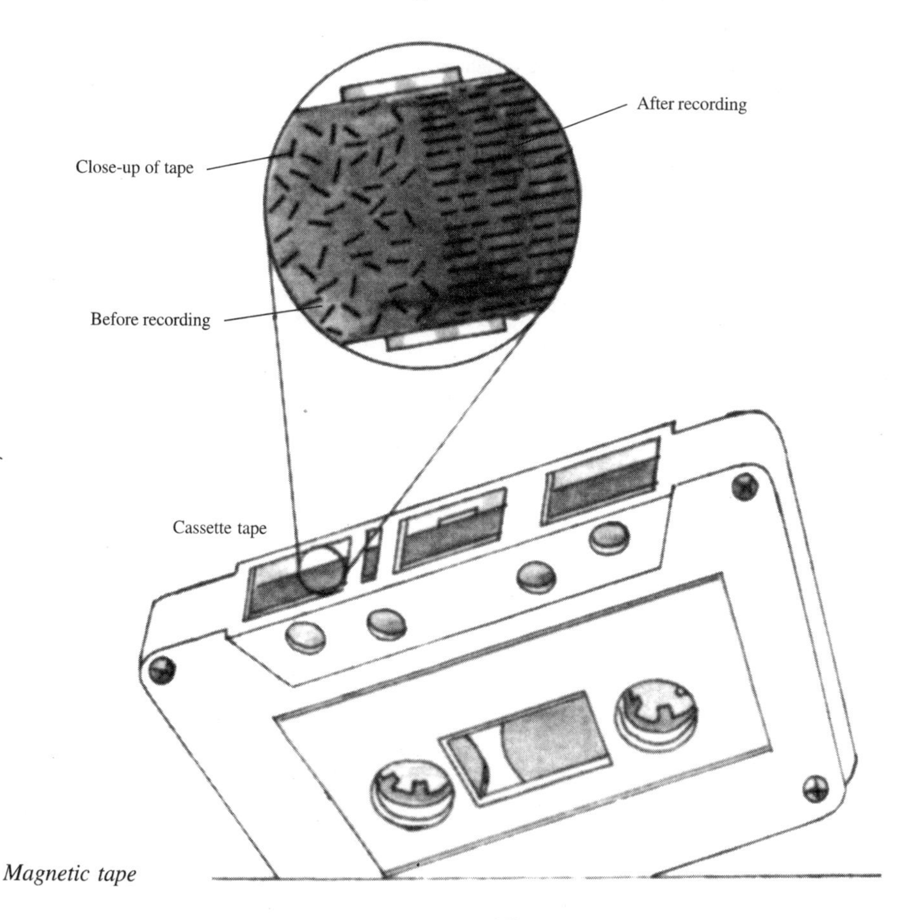

Magnetic tape

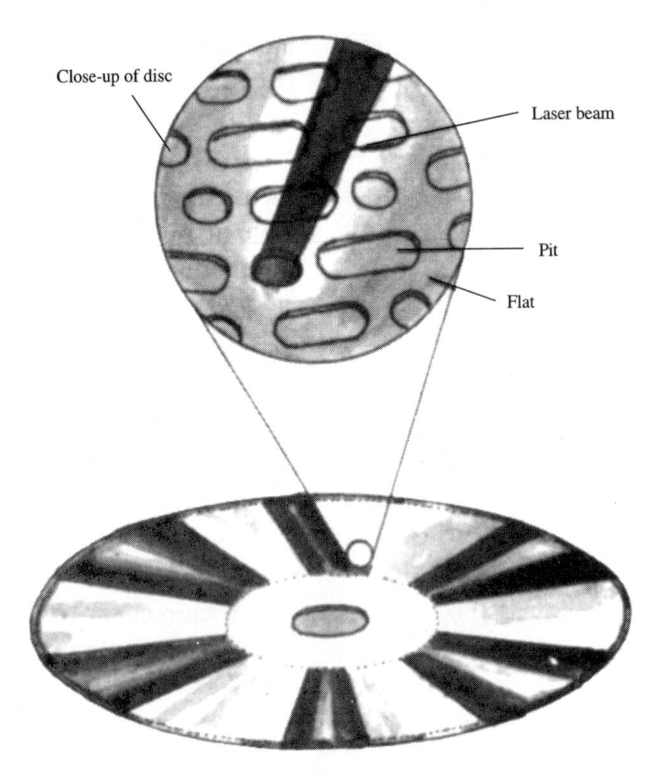

Compact disc

an amplifier and then the original sound is produced by the loudspeaker.

Compact Disc

The electric signals of a microphone are recorded on a compact disc by *laser beam*. Laser is also used for playback. The signals are recorded in the form of digits (in the series of 1 and 0). We have the best reproduction of sound by this method. A compact disc measures only about 12 cm but in comparison to 30 cm-long playing record, its playing time is much more.

■■

TELEPHONE NETWORK

Telephone network is the most effective and useful two-way telecommunication system. With the help of a telephone, people can talk with each other. The use of computers in telephones has added new dimensions to telecommunications. *Viewdata, teleshopping, electronic banking, electronic offices,* etc. are all born out of computers. In fact, computers have added a new chapter to the phone network, which has made communication systems very easy and fast.

How Phone Works

When we speak, our sound creates vibrations in the air. The microphone fitted in the mouthpiece of the phone changes these sound waves into electric current. This current reaches the earpiece of the receiving phone through cable. In the ear-piece of this phone, the electric current again changes to the original sound waves. In this way, your voice reaches the destination. This system is called analog broadcast because in this system,

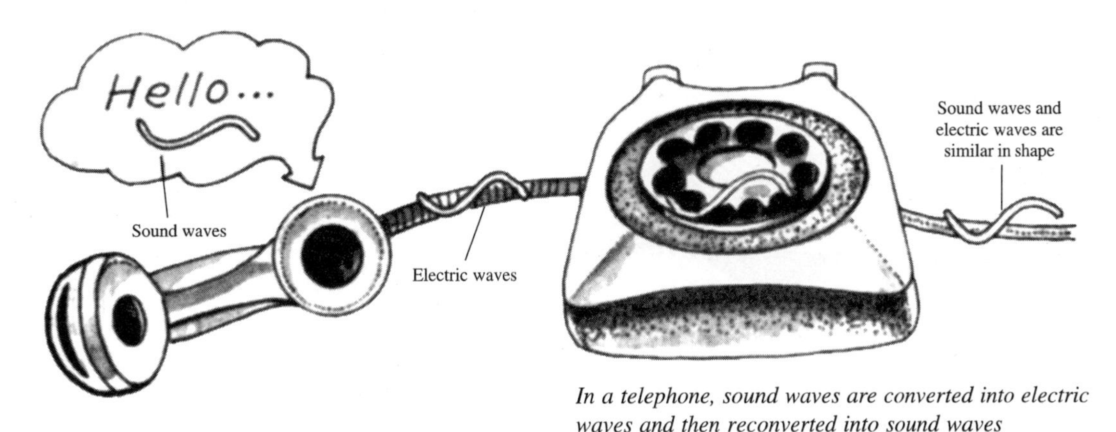

In a telephone, sound waves are converted into electric waves and then reconverted into sound waves

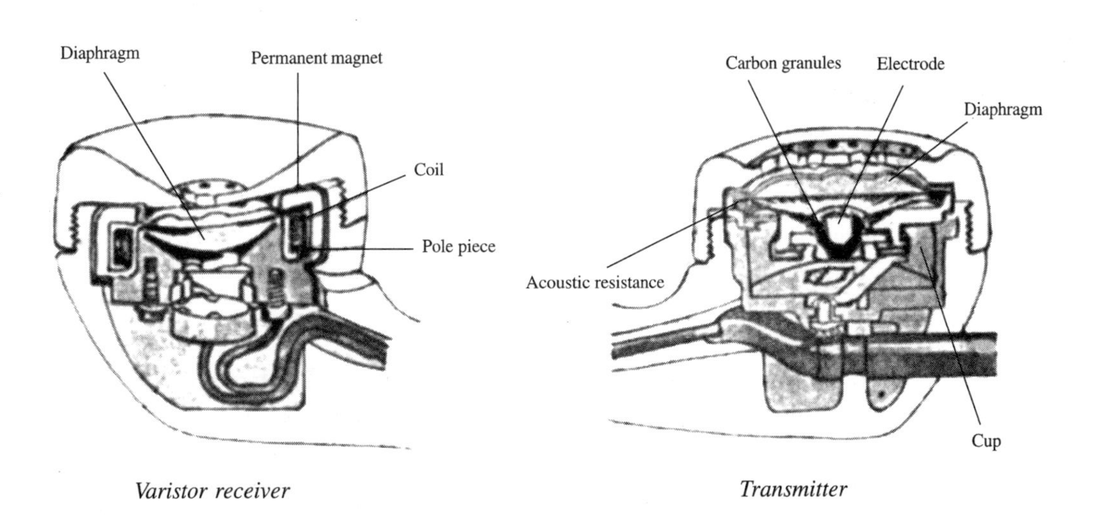

Varistor receiver

Transmitter

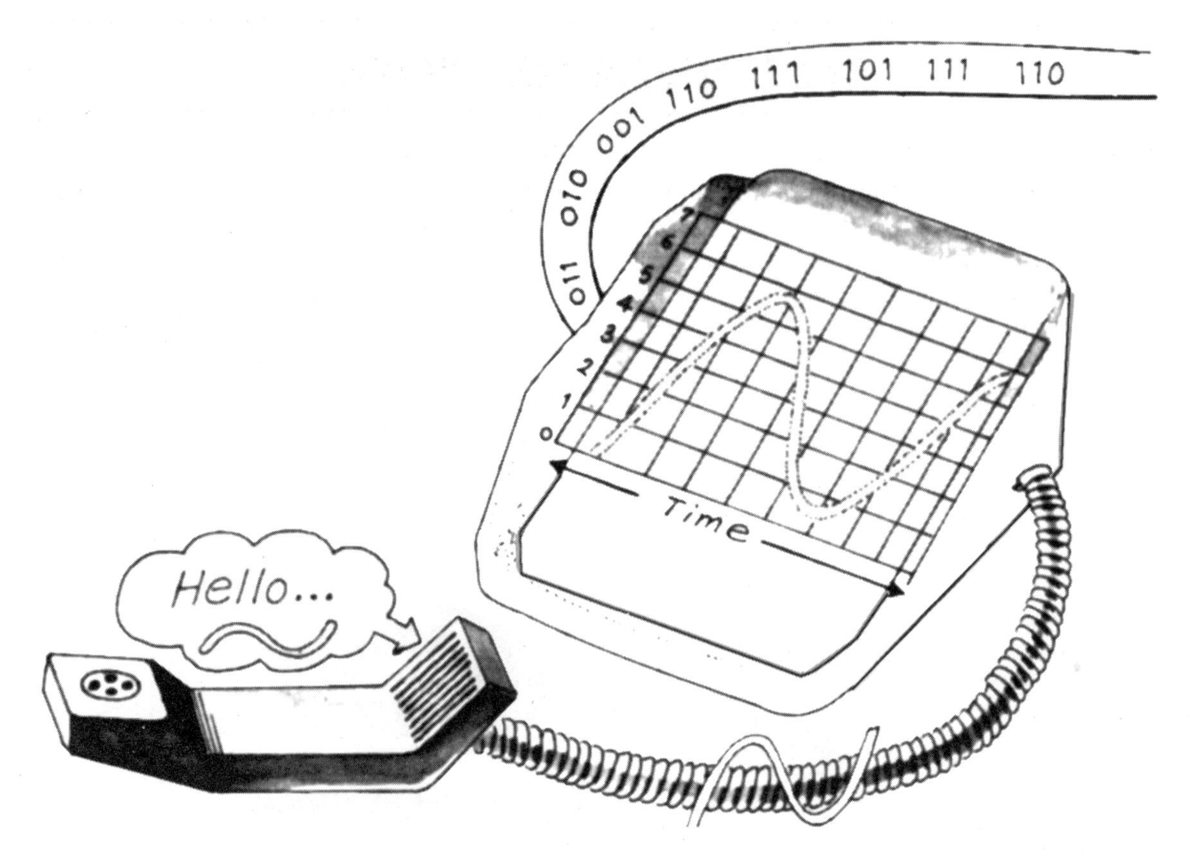

A digital phone firstly converts sound into electric waves. Then these are developed into a number in the binary code. For listening to the sound, the digital signals are converted into the analogue form.

electric current is 'analogous' or identical to sound waves.

Digital Phone System

The *binary digital system* was developed in 1960. In the digital phone system, sound signals are converted into binary digits. Analog signals are converted in the form of 0 and 1 which represent the on and off states. Binary digit is called a *bit*. Seven bits are generally used in telephone communications. The receiving phone converts these digits back into sound waves. Binary digits are fast and are of the same kind, so it is very easy to convert them into their original form. Nowadays, these systems are used on a large scale along with computers.

Fibre Optic Cables

Most of the phone calls travel through copper cables in the form of electric current, but nowadays these calls are also sent through optical fibre cables in the form of light. These cables have proved better than copper cables. Through this cable, about 10,000 telephone calls can be communicated at a time. In this system, laser beam and cable made of optical fibres are used.

Computer Exchange

The telephone system has not been totally digitalized in our country yet, only a few exchanges have been computerized. These exchanges control much more information in less

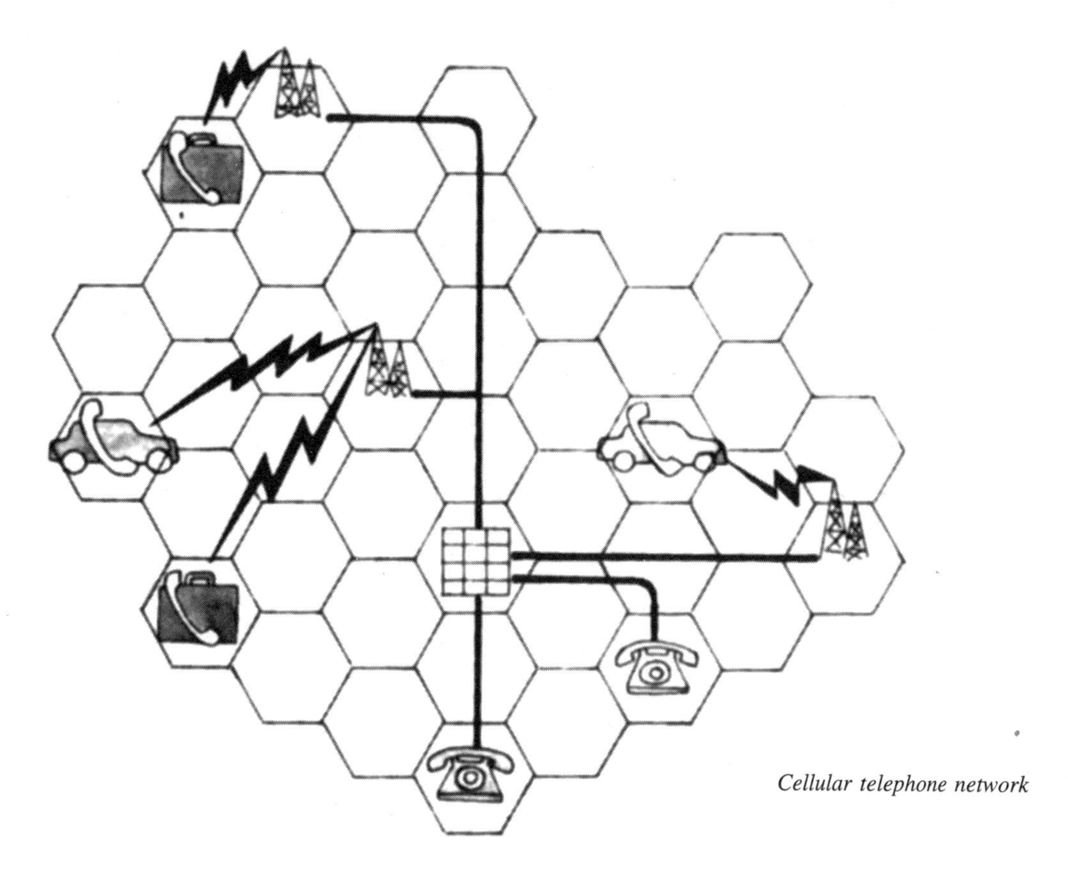

Cellular telephone network

Base station

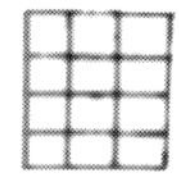

Microwave or landline link

Electrical signals

Car phone

Portable telephone

Private telephone exchange

Public telephone exchange

time in comparison to the old mechanical exchanges. The chances of misconnections, crossed lines, interference or lost calls in these computerized exchanges are extremely low. Such exchanges also provide additional facilities like rerouting of a phone call for another phone, viewdata phone, number directories, automatic monitoring of call, etc.

Cellular Radio and Telephones

Cellular radios or telephones are used for mobile communications. Telephones remain linked with the cables and the person using the cordless phone can use it only within a cell of a few metres from the *base station.* In a few parts of big cities, special types of radio channels or transmitting frequencies are used. In cellular telephone network, an area is divided into many cells. Every cell is at a distance of five kilometres from each other. When a person using a cell

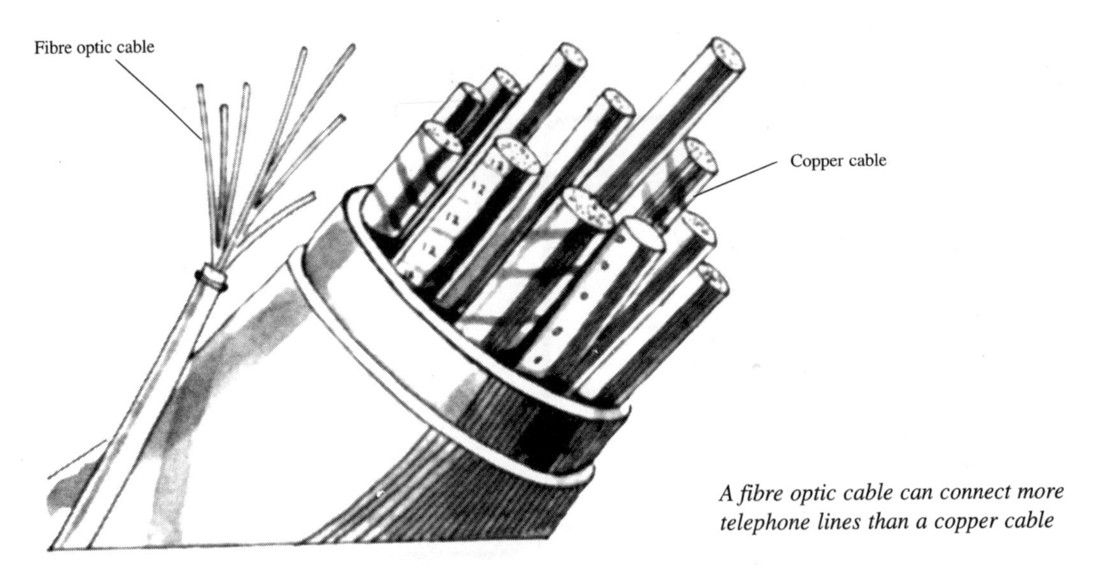

A fibre optic cable can connect more telephone lines than a copper cable

phone makes a call, a radio signal is sent to the nearby base station. The base station sends it to the nearest *mobile telephone exchange*. After that, the call is automatically transferred from the transmitter of one cell to the other cell. In this system, many calls can be made simultaneously.

■■

FAX

Fax or *facsimile transmission* is a new system of sending information on paper through telephone line. It is different from other communication systems like *telex* and *electronic mail.* All types of documents, either printed or handwritten, line diagrams or photographs can be sent or received using the fax machine.

A modern fax machine takes less than 30 seconds to send a written paper. In the beginning, fax machines were of the analog type but now digital machines are in common use.

A fax machine scans the document by light and the image is changed into electrical signals by *photocells*. The message travels through the telephone line and is received by the fax machine at the other end. After decoding, the machine produces a printed copy of the original document.

The coding and decoding system of all the fax machines in the world is the same. So you can send the document from your fax machine to any other fax machine. Every fax machine has a number like a telephone, which has to be dialled before sending a message.

■■

Electric signals are transmitted by the phone

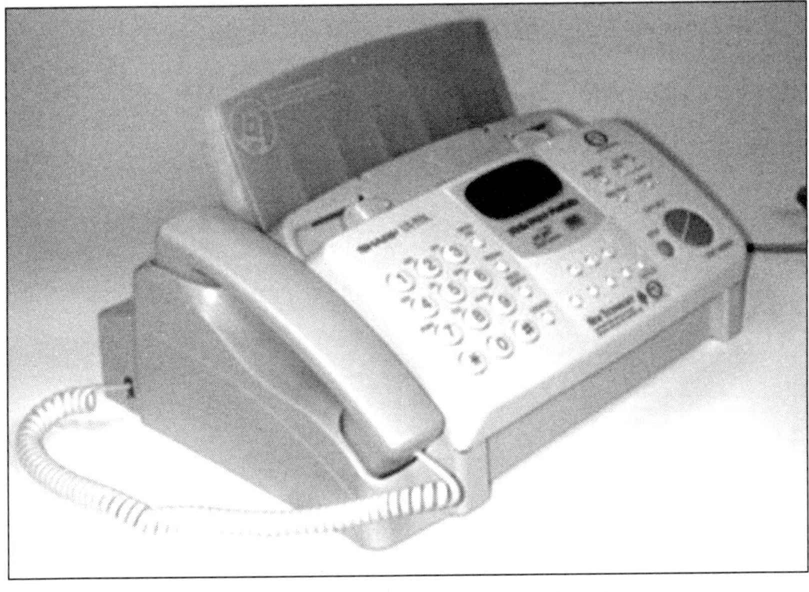

A fax machine can send or receive every sort of word or picture

SATELLITE COMMUNICATIONS

Telecommunication messages are not only broadcast through telephone cables but through satellites also. Everyday, thousands of phone calls, TV signals and computer data are transferred from one part of the world to another through satellites. A *communication satellite* can handle more information than cable and communicate them much faster.

Sending Signals

Messages are sent to the satellite from any earth station in the form of *microwave signals.* Microwaves are a type of short radio waves and can travel through space with the speed of light. Signals are sent and received through dish-shaped antennas. The messages sent from the earth station are received by the satellite and are retransmitted to the earth. The antennas located on earth receive the messages and send them to the destination. These messages can be for telephone, radio or television.

Using Satellites

We can send telephone, radio and television messages to any part of the world with the help of communication satellites very easily. Any event can be telecast live from any part of the globe. Now, it is possible to transmit a telephone call which can travel around the globe and can be received at the same point from where it was transmitted with the help of communication satellites.

Satellites have proved very useful for *multinational companies.* Different offices in the world can be linked to each other. Office work stations have communications between each other.

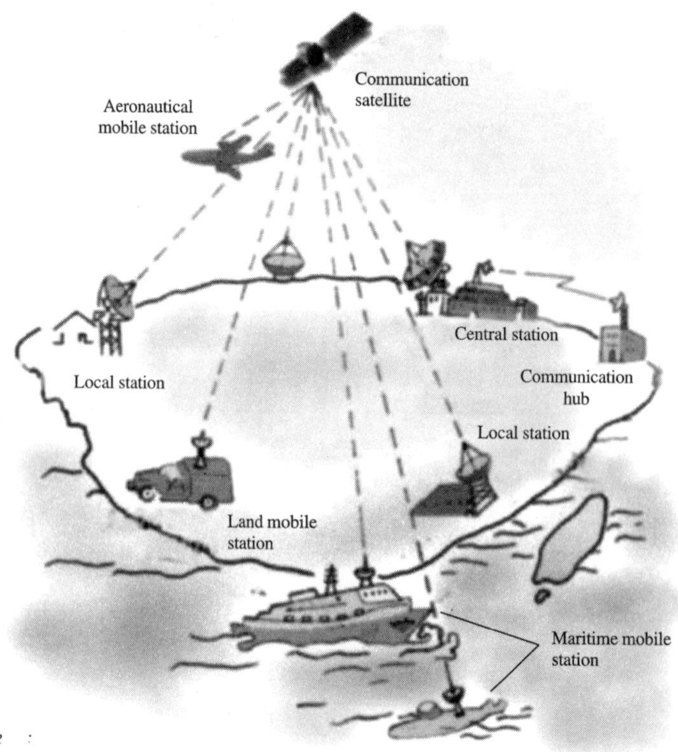

Communication satellite

TELEVISION

The first public demonstration of television was given by John Logie Baird of Britain in 1926. The television which displays the different objects and scenes on the screen is a result of the systematic and joint efforts of the scientists.

V. Zworykin of America made a significant contribution in this field. In 1928, he developed an electronic system, which superceded Baird's mechanical system. It was a revolutionary development for telecasting programmes quickly and correctly.

To telecast TV programmes, sound and scenes are first converted into *electromagnetic waves*. These waves are again converted into sound and scene by the TV set. A TV camera consists of an *orthicon tube*. The image of the scene formed through a lens falls on a photosensitive plate of this tube. Electrons are emitted from the plate according to the intensity of light. After that, it is *scanned by a cathode ray tube*. Scanning changes the image into electric current. It is called a *video signal*. It is telecast after amplitude modulation. In addition to this, sound is converted into electric current by a microphone and is transmitted after frequency modulation. It is called an *audio signal*. These electro-magnetic waves of scenes and sound hit our TV antenna. These are received by the TV set where they are converted back into the original scenes and sound.

The working of colour TV is almost the same as of a black-and-white TV. The light coming from the scene is divided into the three primary colours by three filters fitted in the

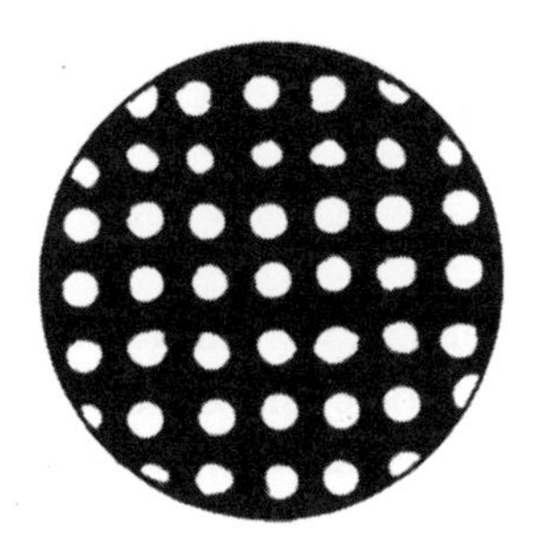

Red, green and blue dots in an image

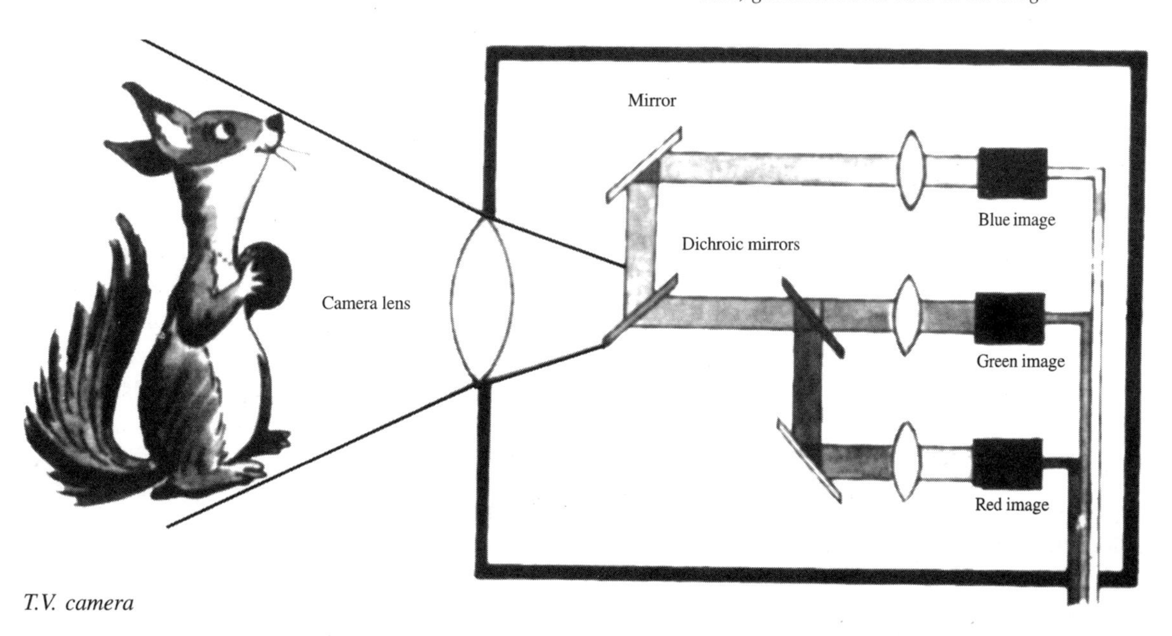

T.V. camera

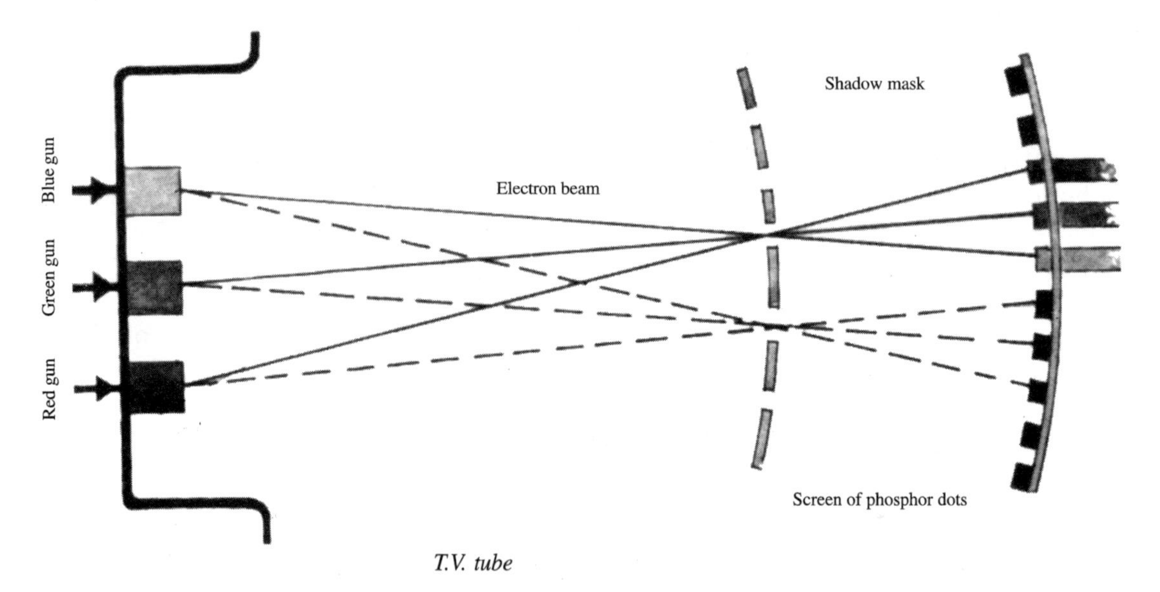

T.V. tube

camera. One filter allows only red colour to pass, the second only blue and the third only green. The light of each colour falls on different camera tubes. Each tube contains a separate glass plate and electron beam. The three signals from the tubes reach the transmitter. The TV transmitter mixes these three signals into one. A black-and-white signal is also mixed with this signal. After that, these signals are sent to the transmitting antenna and telecast. This signal reaches our TV sets. Three electron guns, one for red colour, the second for blue and the third for green are fitted in the TV set. A layer of about 125 lakh phosphor dots of the three colours is painted on the screen of the TV set. These dots are arranged in a pattern of three and emit light when electron beam falls on them. Out of these dots, one emits red light, the second blue and the third green. The colour emitted from each group of dots depends upon the intensity of electron beam. The coloured scene is created on the screen by mixing of these three primary colours in different ratios.

DBS TV

Artificial satellites brought about a revolution in the art of TV transmission. TV stations telecast programmes for different countries with the help of satellites. Nowadays, direct transmission is also done from the satellite. This is called *DBS (Direct Broadcast Satellites).* In this system, your TV set receives signals directly from the satellite. This system requires a special *dish aerial.*

Cable TV

In Cable TV, the signals reach our TV set through a cable. Nowadays, a dish aerial is installed in a

Dish aerial for D.B.S. T.V.

colony and programmes can be viewed by the people of the colony on their TV sets through a cable. More channels can be transmitted through cable than ordinary transmission. TV signals can be sent to hills, valleys and even tall buildings with the help of the cable. Cable TV has become very popular these days.

■■

VIDEO

Video cassette player/recorder

Video recording is a modern technique in which sound and picture are simultaneously recorded on a tape or disc electronically. Television studio makes use of video to record programmes and feature films. Video films can be purchased from the market and can be taken to home to see on one's own TV set. Video tape costs less than ordinary films to make domestic movies. It is also very simple to use them. Disk is the latest development but it is different from the tape.

In 1956, US Ampex Corporation developed the first video recorder. In 1976, Philips Company of Europe manufactured the first video recorder for domestic use. The most popular home video system is *VHS (Video Home System)*.

Video camera

This was developed by JVC in Japan in 1970. In this, a 12.65 mm wide tape is used.

Camera and Recorder

Video camera converts light and sound into electrical signals. Recorder converts these signals into magnetic pulses and records them on magnetic tape.

Video Cassette Player/Recorder

Video cassette player again converts the magnetic pulses which are recorded on tape into the original electrical signals. These signals are again converted into original picture and sound in a TV set. Most of the domestic video cassette recorder machines can record as well as play the tapes. You can record TV programmes on these tapes and watch them afterwards.

Video Disc

Video disc resembles *long play record (Audio LPS)*. You cannot record your own programmes on these discs. They are pre-recorded. A special type of disc player is required to play them.

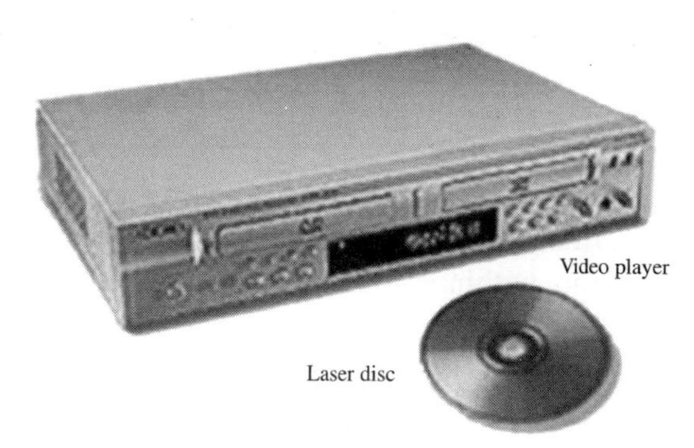

Video player

Video disks are of two types: one is used like audio players and for the other, a thin laser beam is used. Laser disc is like a silver mirror which shows rainbow colours. It is played by *laser beam.*

The programmes recorded on video disc are very clear. They can be preserved for a long time. They are also cheaper in comparison to video tape. These are used in the field of education, training and industries. A few video discs can be controlled by the domestic computer also.

COMPUTER

Computer is an automatic machine which can do calculations, write and solve complex problems within seconds without making any mistake. Not only this but it can also perform many functions at a time. A single computer can do millions of calculations within a second.

Computers are being used successfully in telecommunications, space research, engineering, science, industries, medicine, education, banks, transport, post offices, railways, sports etc.

The first electronic computer was ENIAC, which was invented in 1946 by Eckert and Mauchly of Pennysylvania University. There were 18,000 electron tubes, 70,000 resistors and 10,000 capacitors in it. Each tube was of the size of a small bottle. This computer was controlled by a team of trained operators. In 1950, the transistor replaced the vacuum tube. Nowadays integrated circuits or chips are used in computers which have reduced the size of the computer considerably. Computers are of two types – the first type of computer is called *analog computer* which measures one quantity in terms of the other and the other one is *digital* in which various problems are solved by using digits.

A computer has three main parts –*Input Unit, Central Processing Unit (CPU)*, and *Output*

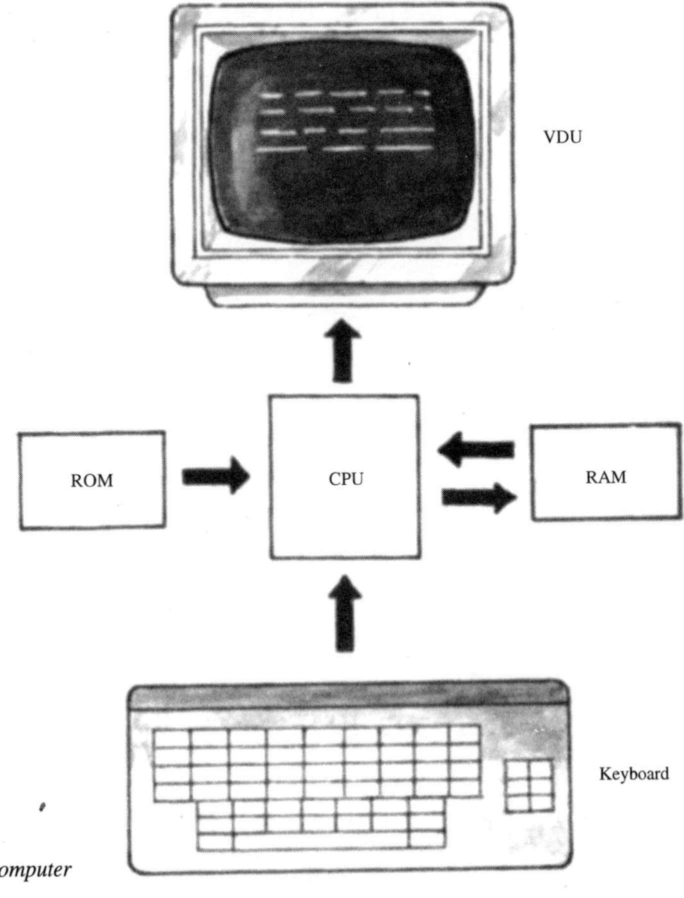

The main parts of a computer

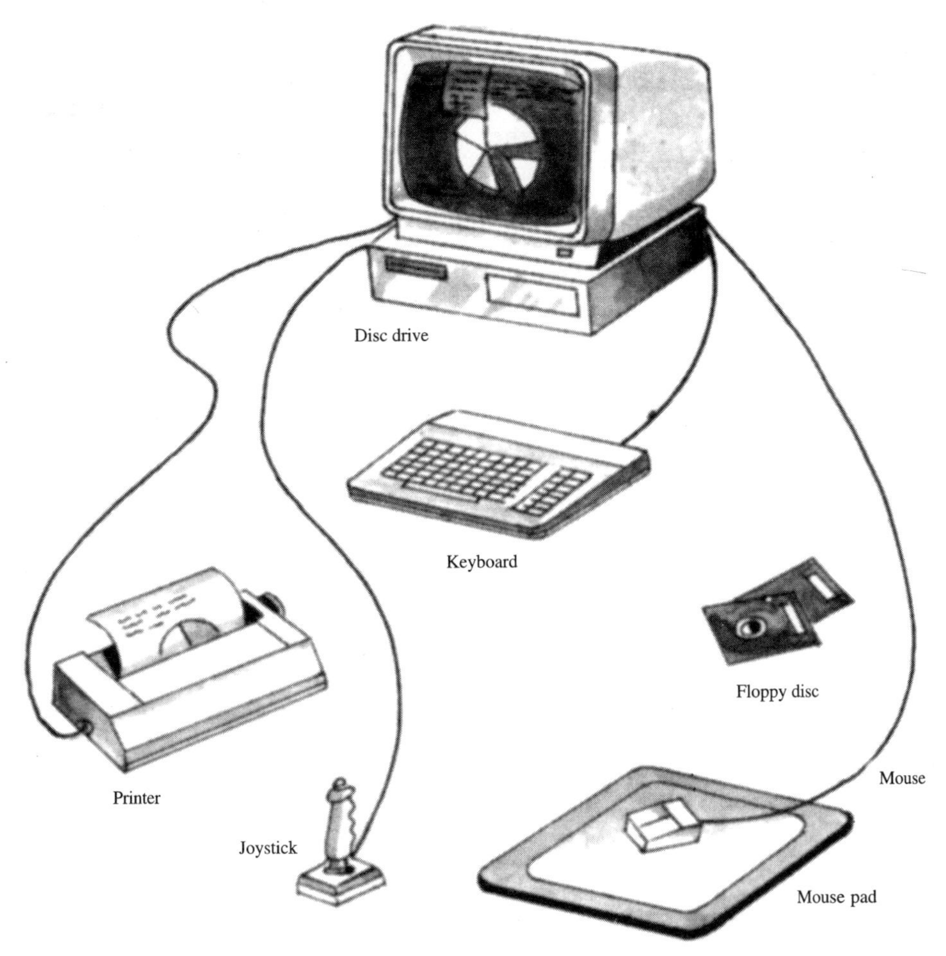

Computer hardware

Unit. In the input unit, magnetic tape, punch card paper or tapes are used. There is a memory in CPU which stores the information. Its controlling system gives directions and arithmetic system performs various operations. Printers, cathode ray tube or sound indicating units are used as output devices. In fact, the input of computer is just like our eyes and ears, CPU functions like our brain and output unit functions like our hand and mouth.

Digital Computers

In digital computers, all information is sent through programme and these bits of information guide the functioning of computer. The data of the programme are sent to the computer through an input machine. This data is processed in the CPU. The data of programme gets stored in the memory. Calculations are done in the arithmetic and logic units. Storage and calculations are controlled by control unit and the solution of the problem is given by the output unit.

Home Computer

The heart of a home computer is only one chip which is called a *microprocessor*. Different parts of a chip perform different functions of the computer. The chip of a home computer has two types of memory, one Central Processing Unit (CPU) and one clock.

Read Only Memory or *ROM* carries necessary messages for the computer. This cannot be changed. *Random Access Memory* or *RAM* is a temporary memory and it is used to give instructions and messages when computer operates. The instructions given to the computer are called a *programme* and messages kept in it are called *data.* Experts wiring a computer programme use special languages which the computer can understand. There are several computer languages like *COBOL, FORTRAN*, etc. but *BASIC (Beginner's All Purpose Symbolic Instruction Code)* is most commonly used.

Computer Software

The programme and data fed to the computer is called *computer software.* We can type words and decimal figures with keyboard which go to Random Access Memory (RAM) of the computer. These do not go into the memory in the form of words and decimal figures because memory cannot store them in that form. Computer changes them into binary code. In this code, all the characters are changed into 1 and 0.

Two binary digits are called *bits.* Most of the computer data are controlled and stored in eight-bit units, which is called a *byte.* The RAM of a single home computer can control about 1,28,000 bytes.

Computer Hardware

The physical parts of a computer which we can touch are called *hardware.* In the main processing unit of a home computer, there is a *Visual Display Unit* (VDU) resembling the TV screen, one *keyboard, mouse, joystick, printer* and *modem.*

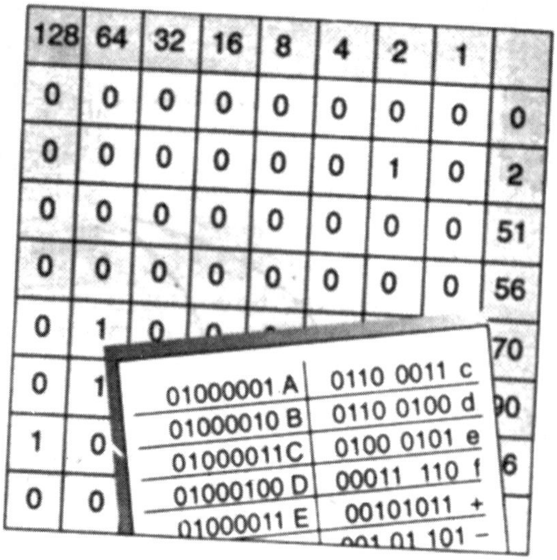

Binary code

Magnetic disks are fitted inside one or two slots of the processing unit. Programmes and data are stored on disks. These are called *floppy disks.*

Mouse is a device which helps you to make a design directly on the VDU. Joysticks are used in video games.

The programme given to the computer can be received in the form of a copy printed by the printer or it can be stored on floppy disk or magnetic tapes. These can be transferred to the other computer through telephone cables. This you can do by modem, which converts computer output signals into audio signals. The modem on the other end reconverts them into original form.

Scientists have made various types of computers today but the fastest and the most powerful among them are *Super Computers.*